LIBERAL ARTS IN JAPAN

LIBERAL ARTS IN JAPAN

Perspectives and Policies in Science and Engineering

Ken Okano

Joshua D. John

Eri Yamamoto

JENNY STANFORD
PUBLISHING

Published by

Jenny Stanford Publishing Pte. Ltd.
101 Thomson Road
#06-01, United Square
Singapore 307591

Email: editorial@jennystanford.com
Web: www.jennystanford.com

British Library Cataloguing-in-Publication Data
A catalogue record for this book is available from the British Library.

Liberal Arts in Japan: Perspectives and Policies in Science and Engineering

ISBN 978-981-4968-05-8 (Paperback)
ISBN 978-1-003-29054-4 (eBook)

Contents

Preface: Education in Japan for the Last 50 Years

The educational system in Japan after World War II was well adapted for rapid increase in population and extremely high economic growth. However, it does not correspond to an aging society with a low birth rate and a drastic change in society. As a result, the present system does not yield internationally competitive students from Japan. Regardless of the physical distance, available time, and economic situations, the educational system in which anyone who wants to learn, can learn, should be introduced through technological innovation.

In most of the universities in Japan, the students are requested to choose their major upon their entrance. Although it might be appropriate for the students who already decided what to study, most of the students cannot decide their major from their interests. In addition, the present system does not allow them either to encounter or to challenge the new field.

In the Japanese educational system, the purpose of the university education was only to increase the specialty. However, the purpose needs to be broadened, including obtaining higher intellectual ability, the support for the late specialized students and/or graduates. In this book, our trial to achieve reform has been introduced.

Ken Okano
Joshua D. John
Eri Yamamoto
February 2022

Chapter 1

What is Education?

What I Learned from My Father

(8 January 2014, speech at my father's (Osamu OKANO) funeral mass)

Let me start with introducing what I have learned from my father.

The first thing I learned is as follows; some of you might agree with me, but my father hardly ever changes his opinion. It sounds as if he was selfish and is very negative, but in fact it seems he was doing his best to allow himself not to change his opinion.

He had been reading four main newspapers of Japan every day, and sometimes he said "I could not sleep well last night and read six books..."

He was trying his best to prove that his opinion was based on evidence and not on emotional feelings, and he always analyzed deeply. This strong will supported him in not changing his opinion.

Second, he always thought about the point of view of the person with whom he was having a discussion. I was very surprised to hear that because he was always very strong and never changed his opinion. His advice was to find the "middle ground" because no one liked to change his or her opinion by force. It might be much easier if one showed the "middle ground" back and forth during the

Liberal Arts in Japan: Perspectives and Policies in Science and Engineering
Ken Okano, Joshua D. John, and Eri Yamamoto
Copyright © 2022 Jenny Stanford Publishing Pte. Ltd.
ISBN 978-981-4968-05-8 (Paperback), 978-1-003-29054-4 (eBook)
www.jennystanford.com

discussion and finished the discussion in a short while having "win–win" feelings.

Third is communicating in both Japanese and English. I am a university professor now, but for the last few years, as you might have heard, the Ministry of Education, Culture, Sports, Science and Technology (MEXT) in Japan has been encouraging educating "global human resources," but I was a bit surprised when I learned that my father followed this criterion.

My father was a bank executive and not an educator, but he realized what would be important in the next 50 years and advised me to master the skills. Although he did not explain to me how he could predict the future, my understanding is that this is set to be my fourth task.

In my 20 years of experience as a university professor, what I learned from him seems to be universal and the essence of education. For me, my father's passing away is really a sad event, but I will continue my struggle to carry out the fourth task for the rest of my life.

What I Learned from My Supervisors

Nature of Education

If we look up the word "education" in the dictionary, it says "Intentional movement towards others in order to change that person in good direction." This matches the general understanding of the word "education" but obviously different from mine, which is "Gratitude to my mentors who brought me up." I am going to explain my definition of "education" in this chapter.

About 40 years ago, I was not really well adapted to the Japanese educational system, the so-called "Jyuken," and was not enjoying my life. What changed my life was again my father's comment: "It might be one of the solutions to revenge 'Jyuken' but also it might be interesting if you could try to be number one in your class."

I could never be number one in my class, but life is interesting, and a rumor spread: "There is a student who is very good at studying and also interested in research. In addition, he has been abroad and can speak English!" This was very rare in Tokai University, which may also be true all over Japan at that time. Many professors offered me to read the additional materials as well as to provide chances to start experiments.

In this situation, I started my research life and met three distinguished professors, which does not include my direct supervisor, and I am explaining what I learned from them based on my memories as they all have already passed away.

- Prof. Tadao Inuzuka
  Former Professor of Aoyama Gakuin University
- Prof. Makoto Kikuchi
  Former Director of SONY Central Research Laboratory
- Prof. Terutaro Nakamura
  Former Professor of the Institute for Solid-State Physics, University of Tokyo

What I Learned from Prof. Tadao Inuzuka

When I was a PhD student, I met Prof. Inuzuka and since then, he assisted me doing some of the experiments and writing manuscripts. The relationship between us was getting worse as my PhD was reaching its final stage. It seemed Prof. Inuzuka felt it and he asked me "Is it possible for you to look after my students at least once a week?" Throughout my visit to his lab every week, I realized Prof. Inuzuka's unique point and charms.

I accidentally found an invitation to submit an article for a scientific journal on his desk when I was doing some experiments with his students. I remembered seeing a similar invitation a few months ago, and I thought "Prof. Inuzuka is asked to write the article regularly" and was very much impressed that he was so famous. But in fact, the invitation is the same one and the second one was just a reminder. I asked him "Is it still in time?" I thought the due date had passed. But his answer was "Listen Ken, if it has already passed the actual deadline, they will never send me a reminder. The fact that I received the reminder tells that we still have time before the deadline." He started writing with a smirk.

I was once asked to assist him in setting up the International Conference site in Sendai with Prof. Sawabe and Dr. Koizumi, both from Inzuka's lab. The task was not that difficult, and everything went smoothly and finished at around 3 pm. My talk was scheduled for the following day, but Prof. Inuzuka had invited us for a beer.

Prof. Sawabe and I tried to find a pub that was open at 3 pm, but not many pubs opened before 5 pm. Eventually, we found one. We drank till midnight, but he made jokes on famous professors in our field and they were really funny. As you can understand, I had no time to prepare for my talk, but my talk was perfect on the following day. What I learned was that "preparation is not compulsory" and since then, I have not prepared for any of my talks.

I was not a graduate of Inuzuka's lab, but he always invited me to his home party in the last week of December, which is called Bounen-kai in Japan. We all promised not to drink too much and stay till very late, but every year, we failed to keep our promise and his wife always gave us a ride to the nearest train station. I still have sweet memories about his home party, but our happiness did not last long as Prof. Inuzuka passed away in a few years. Of course, I attended his funeral, but I remembered that I could not stop crying in front of many people.

What I Learned from Prof. Makoto Kikuchi

Prof. Makoto Kikuchi was a Director of Kikuchi Distinguished Laboratory in Electro-Technical Laboratory and a Director of SONY Central Research Lab, Japan. He has been one of the most well-known researchers in early semiconductor development in Japan.

He was not only known as a physicist, but his soft way of talking to people was always welcome. Sometimes he appeared in TV programs in NHK, and thus, my mother was a big fan of "Makoto Kikuchi."

He loved talking to young people and was so enthusiastic about experiments. He one asked me, when I was a junior researcher, "You are doing something very interesting. Can you explain to me what you are doing?" I started consulting with him since then.

After retiring from his directorship in the SONY Central Research Lab, he still held an office in Shinagawa, and I sometimes visited him for discussion. He obviously liked discussion, and sometimes our discussions made him cancel some of his scheduled meetings.

Once my manuscript was not being accepted for quite a long time, and I consulted with him. "May I see your manuscript and, if possible, experiments?" he asked me. There was nothing to hide, and I showed him everything. His comment was "I could not find any reasons for this manuscript to be rejected for publication.

I will consult with one of my old friends who is now an editor of the journal." As a result, the editor found the appropriate reviewer and he/she accepted my manuscript in a couple of weeks. What I learned from this experience was that even when the science itself is not wrong, there is a possibility that the manuscript will not be accepted, which is largely because of human interaction.

He had been very kind to me, but he also asked me to follow his policy of taking care of young people. He said, "I am happy to discuss with you upon request but don't include my name as the co-authors."

I did not quite understand his intention as I was a young researcher at that time and asked him why. His answered, "I am aged and famous in a sense. If you include me as one of the co-authors, the reviewer might recognize my name, which may not result in a non-biased (fair) decision. In addition, if the invention has a very high impact, the reputation should only belong to real co-authors and not to the people slightly involved."

I was (and am) very much impressed by his comments. He wanted to be one of the scientists, Makoto Kikuchi, and at the same time, he wanted to encourage young researchers like me, as much as possible. In order to achieve both, he imposed himself the rule. I am afraid to say, but I do not think I can reach that level.

What I Learned from Prof. Terutaro Nakamura

Prof. Terutaro Nakamura was a professor of the Institute for Solid-State Physics, University of Tokyo, and he was a well-known specialist of ferroelectrics. He was also Editor-in-Chief of the *Japanese Journal of Applied Physics* and well known for writing Scientific English, as he edited the book called "How to write scientific papers in English" (translated from Japanese). After his retirement from the University of Tokyo, he moved to the Tokai University, and I was appointed as the Research Associate in his lab.

When I first wrote the manuscript in English, he kindly revised all of them, word for word, by confirming my intention. His commented, "As you were the returnee, your English skill is very high and I am interested in supervising your English to top level researchers," and our training started. As my house was quite close to Prof. Nakamura's house, he invited me to his place.

Here, I will explain the typical advice I received from him.

§1. **Introduction**

Recently, much attention has been attracted by the fabrication of diamond using a thermal filament and plasma CVD method under atmospheric pressure.[1-3] As a consequence, the fabrication of semiconductive diamond films for use as electronics components is being investigated. A method for boron doping has been developed which utilizes a gas source such as methane (CH_4) into which diboron (B_2H_6) is mixed.[4,5] Unfortunately, however, B_2H_6 is poisonous.

This is the first report of fabricating semiconductive diamond using boron trioxide (B_2O_3) powder dissolved in organic compounds. Using this method, any poisonous material is dispensable. Furthermore, the present paper is the first to establish definite characterization of B-doped diamond thin films by measuring the resistance and determining the activation energy.

Films are deposited on a Si substrate by thermal filament CVD method using the above mentioned reactant gas. The obtained films have been identified as diamond by means of SEM, RHEED and Raman spectroscopy.

例1 Example 1 (Jpn. J. App. Phy. vol.27, (1988) p.173)

In the second line of this example, there is an expression;

"much attention has been attracted by ..."

Although I had spent a few years in the United Kingdom when I was small, I had never used such an expression before. We do not usually use such expressions in conversation, and it was my first instance of using English to describe scientific phenomena. Later, one of my friends mentioned that she would not use this expression. I fully understand her comment as a native but still believe Prof. Nakamura's advice was fresh to me and really impressing.

His guidance was so kind and warm. I learned many skills and most importantly, how to include my intention or, in other words, eagerness in the manuscript. Even now, when I write English papers, it is usually "accepted as is" in first-class journals largely because of his supervision.

After our English lesson, his wife treated me nicely and her delicious meal always encouraged me to continue doing research. Prof. Nakamura passed away 20 years ago, and since then I do not think I have anything to return to him.

What I Learned from My Parents and Three Mentors

It is obvious that I cannot be a university professor without my parents and three mentors. I once asked my mentors, individually, "You are so kind to me. All your advice is valuable, and I really appreciate it, but there is nothing I can return to you at present as I am only a young researcher. Can you advise me how I can return to you?"

It was surprising that all of them answered to me with the same intention: "There is no need for you to return the gratitude to me. I, myself, was also educated by my mentors, and I couldn't return to them even until now. Please don't forget to 'educate' young people who might need your assistance as a return of your gratitude to me."

I was much impressed with their life policy as well as words. The nature of education has nothing to "do something to people" but "to return what I have been supervised"; this is what I learned from three professors.

What I Want to Do for My Students

Compared to my mentors, my name is not so big and thus I believe I cannot do many things. Even in this situation, every day I am trying to support my students as much as possible.

Compared to my student period, modern students are smarter and more sensible. It seems some of them are not taught that their smartness is one of the powerful tools for "studying" but not for "research."

I sometimes ask my students, "What is the difference between 'studying' and 'research'?"

They often have no clear distinction between the two, but I have a clear vision of the difference. For me, studying is "to learn (or memorize) what is written in the textbook," while research is "to find or think what is not written in the textbook."

It may not be too much to say that we are using different parts of the brain unintentionally for "studying" and "research," which suggests that the talent for studying is not always effective for research. In other words, the students who are not good at studying might be a first-class researcher in 10 years. I will try not to forget my conclusion and continue supervising my students.

Chapter 2

What We Have Been Doing in ICU

My Talk in ICMAT, Singapore, 23–28 June 2019

Thank you very much again for staying till the very end of this session today. My name is Ken Okano, and I will give a talk on Liberal Arts education for Science and Engineering students in Japan.

To be honest, I am not an educator at all. My hobby is doing experiments and research. I think, in Japan, we are a leading country in technology. But I do not think the education part is working properly, so I am going to propose the idea and I would like to hear your opinions after. No science at all, so feel relaxed. You can also interrupt me if you have any questions. These slides are written in both Japanese and English. Here, what I am trying to say is that we need to find the future direction of Japanese education as soon as possible. It is one of the biggest tasks of Japan because as you know, education usually takes a lot of time, even compared to research. I mean research is not that difficult. But for education, it takes more than 10 and 15 years. So we should think about it in advance, and we are at least focusing on education for the variety of society. Which means that after the war, there has been an educational system in Japan. From my point of view, the main purpose of that system is to train people, because we do not have anything to eat and there was no point in doing education. And now, what we should do is to

Liberal Arts in Japan: Perspectives and Policies in Science and Engineering
Ken Okano, Joshua D. John, and Eri Yamamoto
Copyright © 2022 Jenny Stanford Publishing Pte. Ltd.
ISBN 978-981-4968-05-8 (Paperback), 978-1-003-29054-4 (eBook)
www.jennystanford.com

change the system. But Japanese people are not flexible, so we are still using the same system that we proposed 50–60 years ago. Can you believe it?

As I said, I was born in Japan and raised in the United Kingdom when I was a child. After I got my PhD in Japan, I was in MIT and the University of Cambridge, and I am still working with them at the National University of Singapore. So I like research.

To start with, this is what I did a long time ago. My research topic is growing diamonds, and especially this is called a blue diamond. Compared to ordinary diamonds in terms of transparency, the blue diamond is very expensive.

And this is very easy to grow, and up until my record in 1987, people had used the gas called diborane, which is very poisonous. Thus, not many people wanted to use it. But in my research, I used boron trioxide instead, and we could make the blue diamond like this, and we could change and control its color from the black color to a very pale blue color (Fig. 2.1). This was my start, and after that, I made the p-n junction diode using diamonds and it appeared in *Nature* (Fig. 2.2).

In addition, I found a strange phenomenon, which is an electron field emission from diamond surfaces (Fig. 2.3). This was also published in *Nature*. I got quite a lot of money, and I think it goes up to $10 billion in total, so it is kind of my hobby to do research and experiment (Fig. 2.4).

But one of the bad things with me is that I do not trust students. So when students say "Professor, I get this result," I would say "You did. Good job." But I do not usually trust them, which is not good for a supervisor, so I have to do it. Anyway, I have received a lot of money, and I am happy doing that.

Let me move on to the introduction of my university in Japan (Fig. 2.5). It is called the International Christian University (ICU), which is a very small private university located in Tokyo and founded in 1949, just after the war. The total number of students is just less than 3000, including those of graduate schools.

One of the advantages of ICU is that the ratio of faculty members to students is quite high. This ratio is one of the highest in Japanese private schools. So one professor can teach a very limited number of students for most lectures.

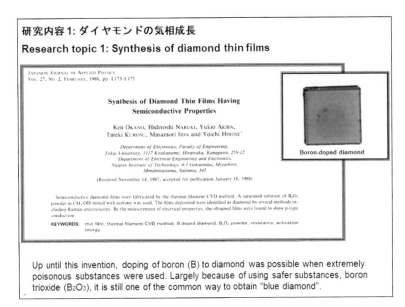

研究内容 1: ダイヤモンドの気相成長

Research topic 1: Synthesis of diamond thin films

Up until this invention, doping of boron (B) to diamond was possible when extremely poisonous substances were used. Largely because of using safer substances, boron trioxide (B_2O_3), it is still one of the common way to obtain "blue diamond".

Figure 2.1   Synthesis of diamond thin films.

研究内容 2: ダイヤモンドのpn接合ダイオード

Research topic 2: Diamond pn junction diode

Everyone believed that n-type diamond cannot be grown, but this report implied the possibility of growing n-type diamond using Phosphorus Pentoxide (P_2O_5).

Figure 2.2   Diamond p-n junction diode.

研究内容3: ダイヤモンドからの電界電子放出

Research topic 3: Electron field emission from diamond

Low-threshold cold cathodes made of nitrogen-doped chemical-vapour-deposited diamond

Ken Okano*, Satoshi Koizumi†, S. Ravi P. Silva‡ & Gehan A. J. Amaratunga§

* Department of Electronics, Tokai University, 1117 Kitakaname, Hiratsuka, Kanagawa 259-12, Japan
† National Institute for Research in Inorganic Materials, 1-1 Namiki, Tsukuba, Ibaraki 305, Japan
‡ Department of Electronic and Electrical Engineering, University of Surrey, Guildford, Surrey GU2 5XH, UK
§ Department of Electrical Engineering and Electronics University of Liverpool, Brownlow Hill, Liverpool L69 3BX, UK

BECAUSE diamond surfaces terminated with hydrogen have a negative electron affinity[1-4] (the conduction band minimum lies below the vacuum level), they are expected to emit electrons spontaneously. This has led to attempts to develop 'cold cathodes'—miniaturized vacuum diodes that might have applications in microelectronics and flat-panel displays. In previous studies of electron emission from diamond grown by chemical vapour deposition[5-9] (CVD), the threshold voltages for emission were more than an order of magnitude too large for use in battery-driven cold cathodes. Although low-threshold emission from caesium-coated, nitrogen-doped high-pressure synthetic diamond was reported recently[9], ultimately diamond thin films grown by chemical vapour deposition (CVD) look to be the most promising material for cold-cathode applications. But to obtain low-threshold emission, it is necessary to introduce high concentrations of donor dopants such as nitrogen—something that is difficult for CVD diamond. Here we report that high concentrations of nitrogen can be incorporated into diamond films by using urea as the gaseous nitrogen source, and that such heavily doped films shown very-low-threshold electron emission, which augurs well for cold-cathode technology.

Generally it is necessary to apply few thousands volt to extract electrons, but the electrons can be extracted from nitrogen-doped diamond only by applying few volts.

Figure 2.3 Electron field emission from diamond.

獲得した主な研究助成金

Major Research Grants (selected):

MEXT Grants-in-Aid for Scientific Research (*Kakenhi*)
(A)#13305006 2001-2004 (P.I.)
予算規模: 3600万円/ Budget: 300k USD

文科省私学高度推進化事業・学術フロンティアプロジェクト 2003-2007
Academic Frontier Project for Private Universities
予算規模: 2億2000万円/ Budget: 1.8M USD

私立大学戦略的研究基盤形成支援事業 (#S0801012) 2008-2012 (P.I.)
Support Program for Private Universities
予算規模: 1億7000万円/ Budget: 1.4M USD

Patents filed:
14 Japanese Patents
2 US Patents

Figure 2.4 Major research grants.

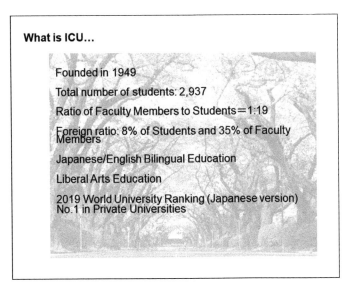

What is ICU...

Founded in 1949

Total number of students: 2,937

Ratio of Faculty Members to Students＝1:19

Foreign ratio: 8% of Students and 35% of Faculty Members

Japanese/English Bilingual Education

Liberal Arts Education

2019 World University Ranking (Japanese version) No.1 in Private Universities

Figure 2.5   What is ICU?

In Japan, we still have a large number of Japanese faculty members and students in ICU: We have 8% of foreign students and 35% of foreign faculty members, which is a very high in ratio. And we have a very strong English training program and the Liberal Arts education, which is very famous in the state but not in Japan, and nor in Europe. In 2019's World University Ranking (Japanese version), we got the first rank for private universities.

Here are pictures of ICU. Since its name International Christian University, we have the chapel. And this is the graduation ceremony.

Are you familiar with liberal arts? Let me just give you a few words for that: education for making people free with critical and creative thinking (Fig. 2.6).

Liberal arts originally derived from the writers. So, liberal arts is very famous in the United States. This is what I got from Harvard University, so as I said, through liberal arts, students gain the ability to think creatively, critically, and independently. That is one of the most important parts of liberal arts.

And this site (Fig. 2.7) was made by one of my former students. What we are trying to say is that since we do not have a fixed program or curriculum, students can choose whatever color, style, and size of shirts they like. This is our ideal program style. It sounds nice, but if you really want to do it in science education, it is sometimes very difficult.

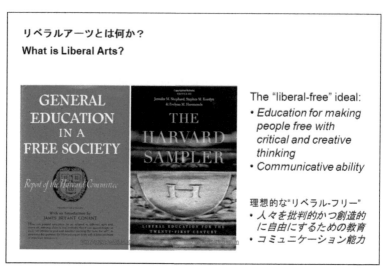

Figure 2.6 What is liberal arts?

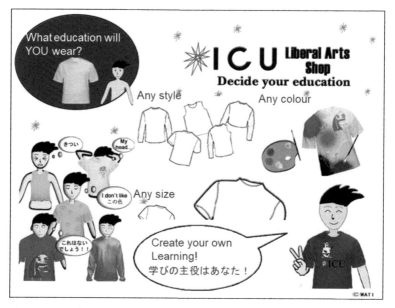

Figure 2.7 What education will you wear?

Before I moved to the ICU, my impression toward ICU in terms of students' graduation and career had been this: There are a number of students who get a job as a journalist and a patent attorney. Can you imagine why they prefer to be journalists and patent attorneys? In those jobs, they do not have to do the experiment or produce something by themselves. They just listen to the story and write down in the way they understand. Of course, they are very creative jobs, and I do not mean that I hate those jobs. But as a scientist and experimentalist, I do not really like those. So after I moved to my university, I forced all the students to do something by themselves, and one of the topics is called the Black Box experiment, which is quite famous in the ICU (Fig. 2.8).

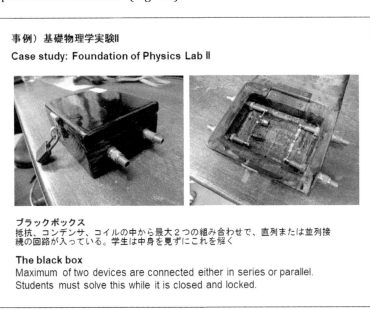

事例）基礎物理学実験II
Case study: Foundation of Physics Lab II

ブラックボックス
抵抗、コンデンサ、コイルの中から最大2つの組み合わせで、直列または並列接続の回路が入っている。学生は中身を見ずにこれを解く

The black box
Maximum of two devices are connected either in series or parallel. Students must solve this while it is closed and locked.

Figure 2.8 The black box.

We have some resistors, capacitors, and inductors inside and students cannot see. Through the several classes, they need to find out what is inside without opening it. This is a kind of puzzle, but I think this is a very nice piece of experiment. I also force students to make a radio (Fig. 2.9) by themselves, and if I cannot hear the music or the voice coming out of the radio, I do not give them the pass grade.

事例）基礎物理学実験II

Case study: Foundation of Physics Lab II

ラジオ制作
授業で得た知識を用いてラジオを鳴らす。鳴らすことができなければ不合格。

Building your own radio receiver
Utilizing the knowledge acquired in the lab, the student must build a radio and successfully operate it. Failure to comply, will result as an E.

Figure 2.9   Building own radio receiver.

As a result, though not everyone, this student presented her research topic in one of the International Conferences and wrote a paper when she was in the third year (Fig. 2.10).

学部生による研究論文

Publication by undergraduate students

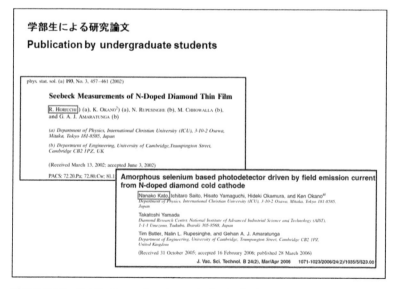

Figure 2.10   Publications of undergraduate students.

Thus, as an output, I think some of them are quite good, and this student wrote two papers before her graduation. This student got a job in SONY after undergraduate degree. In this way, we have good examples even after those kinds of not very research-oriented programs; some of them find a way to go.

Figure 2.11 shows an old-style thickness monitor that is broken in my lab. When this was broken, since we do not have an engineer, we could not do anything. Instead of repairing, a company sent a quote for the present model, which costs 2 million yen. I did not want to pay that, so I just offered one of the students and we fixed it. He said that even though he was not a very professional student, he really liked it. So I said that if he succeeded in making one of those, I would give him the pass grade. He said "okay, I'll do it." Within a couple of weeks, he made it and his invention saved me a lot of money (Fig. 2.12). Additionally, from his point of view, he did something.

自分たちの手で...
Using our own hands...
壊れちゃったけど、あまりにも古いので修理はできず.
代替品は200万円！　🤑🤑🤑

水晶振動子膜厚計 (メーカー製)
Crystal oscillator thickness monitor

Figure 2.11 Crystal oscillator thickness monitor.

Now I offer a kind of model lecture to high school and junior high school students. What I have found in Japan is that most of the students are quite good at studying in terms of learning and memorizing what is written in the text. However, if you just continue studying, we never reach the goal of research. From my understanding, research is to find or think what is not written in the textbooks. So I try to use these kinds of slides in order to encourage young students. But even adults do not know the difference between studying and research, so I was a little shocked. I have been doing my best to tell the difference to students as well as teachers. This point is quite difficult to tell to junior high school and high school students in Japan.

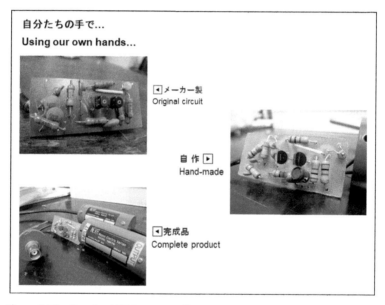

Figure 2.12 Repaired thickness monitor.

Although the total number of students who graduated from my lab in the ICU is very limited, we have 10 PhD holders in total. I have been teaching in the ICU for 20 years, so it is not very bad. What is more, people outside of the ICU have given very nice comments and they all like ICU's graduated students. I think we are doing a good job in educating future researchers, not only those who decide to have a job afterward. What I am trying to say is that in most of the Japanese universities, students need to decide their majors before they enter. This system works well for those who recognize their interests, but may not be the best for those who want to find some new interest or to try something new after their entrance. The original purpose of education is not only to pursue a specialty but to provide intellectual findings for all of those who really want to learn. We need to reconsider the Japanese educational system. As I said, the old Japanese educational system is training oriented, and there seems to be no real education. Therefore, we have to engage the intellectual part rather than just training. That is the future goal. Thank you very much.

Chapter 3

Tips for Future Education

Foundation of Physics (PHY103)

Joining the Course

This is an introductory course on physics. It typically resembles a first-year course for physics major students in typical universities. However, at the ICU, due to the liberal arts approach, students are free to take the course at any stage until their graduation.

To facilitate the students in planning their course credits, the course syllabus is available throughout the year on the school website (Fig. 3.1).

The syllabus gives a general description of the course, the aims, and the objectives. The course topics and schedule are given along with the languages of instruction and communication. The grading policy, expected study time, and references are also given. With this information, the students can decide whether the course aligns with their objectives and scheduling. If they are interested, they can register for the course during the registration period. Even after the registration period, there is an additional late registration and registration change period where students can make changes to leave or join the course.

Liberal Arts in Japan: Perspectives and Policies in Science and Engineering
Ken Okano, Joshua D. John, and Eri Yamamoto
Copyright © 2022 Jenny Stanford Publishing Pte. Ltd.
ISBN 978-981-4968-05-8 (Paperback), 978-1-003-29054-4 (eBook)
www.jennystanford.com

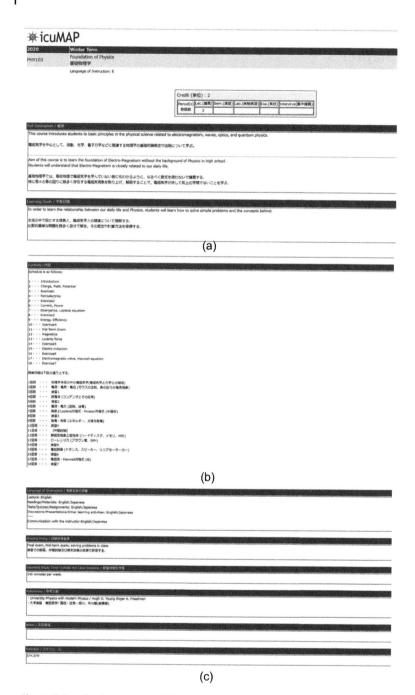

Figure 3.1 Physics course syllabus.

At the registration, the lecturer is provided with an enrolment list showing the students who have joined the course. There is also no restriction by major; any student is accepted. A typical distribution is shown in Fig. 3.2.

31265   PHY103   as of December 4, 2018 7:24 PM

Course Name	: Foundation of Physics	
Schedule	: 3/M,3/W	
Instructor	: OKANO, KEN	T_Enroll : 38

single major: AS(maj1), double major: AS(maj1,maj2), major minor: AS(maj1/maj2)
** = Students who are expecting to graduate in this term.
✳ = Auditors, Credit Auditors (*kamoku-to rishusei*), Special Exchange Students, etc.

G	STID	NAME_E	NAME_KJ	DIVISION
**	19▅	▅▅▅▅▅	▅▅	AS (PHY)
**	19▅	▅▅▅▅▅	▅▅	AS (PSY)
	19▅	▅▅▅▅▅	▅▅	
**	19▅	▅▅▅▅	▅▅	AS (PHY)
	20▅	▅▅▅▅	▅▅	AS (PHY)
	20▅	▅▅▅▅	▅▅	AS (BIO/LIT)
	20▅	▅▅▅▅	▅▅	AS (BIO/CHM)
	20▅	▅▅▅▅	▅▅	AS (PHY.LED)
	21▅	▅▅▅▅	▅▅	
	21▅	▅▅▅	▅▅	
	21▅	▅▅▅	▅▅	
	21▅	▅▅▅▅	▅▅	
	22▅	▅▅▅	▅▅	
	22▅	▅▅▅▅	▅▅	
	22▅	▅▅▅▅	▅▅	

Figure 3.2   Enrollment list of students.

Here we can see that for students who have already declared their major, there are students from physics, psychology, biology, language education, and some minors in literature and chemistry. Based on the student IDs, we can see graduating (prefix 19), third-year (prefix 20), second-year (prefix 21), and first-year students (prefix 22). As a result, the spectrum in terms of background for the students enrolled tends to be very wide: from final-year students to first-year, first-term students, from physics majors to economics or humanities majors. For this reason, the course is designed to have flexibility so that the material is relevant to a generally wide student base.

At the beginning of the term, the lecturer and teaching assistant provide several informative materials for the course. The class schedule is given both as hard copy and online in the class Moodle (Fig. 3.3). All the registered students are requested to enroll to the course Moodle and are provided with a secret access key. This class Moodle is used for providing course resources such as lecture slides, tutorial materials, and references among other things. The resources remain accessible throughout the term, and until the students graduate.

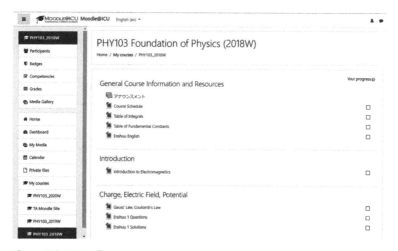

Figure 3.3   Moodle course.

Taking the Course

Lectures

The lecture slides are mostly in English. The lecture itself is delivered in English. If there are students with questions, they can ask in either

English or Japanese, whichever they are comfortable with. The lecture will respond to the questions in both English and Japanese to facilitate better understanding for all the students. Figures 3.4 and 3.11 show a typical lecture in slides.

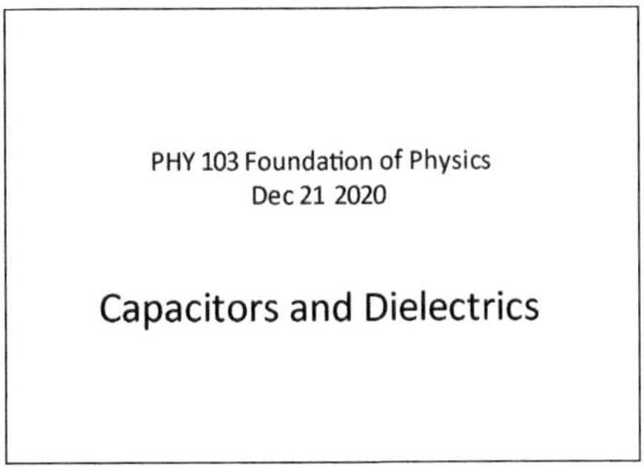

PHY 103 Foundation of Physics
Dec 21 2020

Capacitors and Dielectrics

Figure 3.4   Capacitors and dielectrics.

(Figure 3.4) Today's topic is "Capacitors and Dielectrics." You may be familiar with the term "Capacitance" but may not be with "Dielectrics." In this lecture, I am trying to explain "Capacitors and Dielectrics" as one of the topics in university physics, which is different from high school physics.

But do not be scared; at least, you all can understand what they are and where they are used in our daily life.

(Figure 3.5) For those who have learned physics in high school, capacitors are one of the popular electric devices in which the electric charges are stored. The schematics are given as upper-right figure, and this is very common for most of you, but have you seen the actual devices shown as upper-left figure?

Also, in high schools, you are forced to memorize the relation between capacitance (C), electric charges (Q), and voltage (V), as shown in lower-left equation. But in this lecture, I am going to show you this equation can be derived from Gauss' law as we learned a week ago.

(Figure 3.6) In this slide, you can see the schematics of capacitance in the upper-left figure. We will apply Gauss' law to the upper electrode, which is in green color. When we think Gauss' closed surface to be a rectangular parallelepiped having the bottom area of A, which is equal to the area of the electrode, the double integral part of Gauss' law can be simply rewritten as "A." Thus, the electric field E can be obtained as $E = q/\varepsilon A$.

Capacitors

◯ What is **CAPACITOR**?

An element that accumulates electric charges

 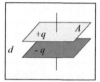

Electric charges are stored between two electrodes (anode and cathode)

⬤ **Capacitance**

$$C = \frac{Q}{V} \quad [F] = [{}^{C}\!/{}_{V}]$$

When a capacitor stores 1C with applied voltage of 1V, the capacitance is defined as 1F (**farad**).

Figure 3.5 What is a capacitor?

Capacitance

⬤ Capacitance: Amount of electric charges stored when 1V is applied

$$\iint_{S} \mathbf{E} \cdot d\mathbf{S} = |\mathbf{E}| \cdot \iint d\mathbf{S} = |\mathbf{E}| \cdot A = q/\varepsilon_0$$

$$|\mathbf{E}| = \frac{q}{\varepsilon_0 A} = \frac{\sigma}{\varepsilon_0} \quad (\sigma : \text{Charge density})$$

In this case, different kinds of electric charges (plus and minus) are stored between two electrodes
-Between two electrodes, the magnitude of electric field is constant at any point
-At rest of the area, the magnitude of electric field is zero
Difference of electric potential (voltage) between electrodes can be expressed as :

$$V = E \cdot d = \frac{qd}{\varepsilon_0 A}$$

From the definition $C = Q/V$, $\qquad C = \varepsilon_0 \frac{A}{d}$

(capacitance of a flat plate capacitor)

Figure 3.6 Capacitance.

According to the relationship between voltage (V), electric field (E), and distance (d), V can be written as $V = qd/\varepsilon A$. Also from the definition of capacitance (C), charge (Q), and voltage (V), we can obtain the well-known relationship $C = \varepsilon A/d$.

(Figure 3.7) This slide shows the capacitors connected in series and parallel. Most of the students learned these relations in high school. But the point here is we do not need to memorize these. For example, in parallel connection, it is obvious that the applied voltage of all the capacitors is the same as $V[V]$. Because $C[F]$ is constant as long as the same capacitors are used, the total charge that can be stored is $Q[C]$ multiplied by n (number of capacitors). In contrast, when we think about series connection, the charges stored in each capacitor is the same. As the applied voltages in total is $V[V]$, the applied voltage on each capacitor is $V[V]/n$.

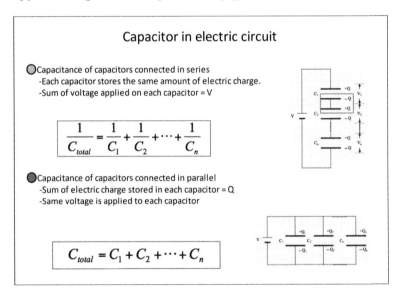

Figure 3.7 Capacitor in electric circuit.

(Figure 3.8) The next slide shows the features of "Dielectrics." A dielectric is a special kind of insulator. Why is it special? Because when we apply a voltage or electric field, a dielectric is polarized.

I am sure you have not heard the word "polarized." Polarization is one of the physical phenomena that can be explained by the localization of carriers in the insulators. Although polarized carriers cannot move freely like free carriers in the metal, a polarized carrier is able to transport electricity.

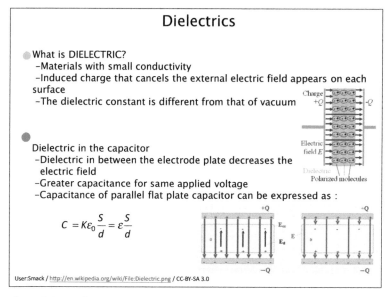

Figure 3.8   Dielectrics.

On inserting dielectrics in the two electrodes of the capacitor, ferroelectrics act as an electric field canceller, so that the internal field is reduced and thus prevents the breaking of the capacitor due to huge voltage.

(Figure 3.9) You may think that dielectrics have nothing to do in our daily life, but in fact, they are used widely in consumer electronics. The upper figure of this slide shows the schematic structure of a capacitor having a plastic sheet as the dielectrics. The role of the plastic sheet is easy to understand. If we go back to the capacitor equation in Fig. 3.5, it is obvious that we need a large A (area) and a small d (thickness) in order to obtain larger C, and this structure is one of the typical examples. In the latest technology, instead of using plastic sheets, we use aluminum (Al) plates as the electrodes with an extremely thin chemically oxidized layer of Al as the dielectrics.

One of the applications is the condenser microphone in which the thin film situated between the electrodes can be vibrated as the sound comes in. According to the vibration, the capacitance changes and we can detect the sound.

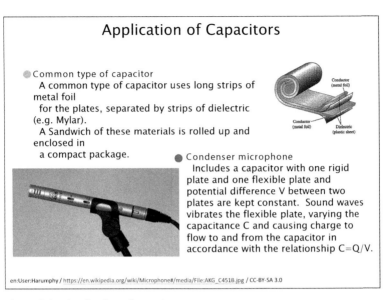

Figure 3.9 Application of capacitors.

(Figure 3.10) The recent application is the touchscreen of smartphones. Our smartphones usually have glass plates on top, which prevent the flow of DC (direct current). But it allows the flow of AC (alternating current) as we observe in the capacitors. The change in the capacitance can be detected, and the smartphone can learn the user's touches.

Let us work out an exercise on calculating the capacitance of a parallel-plate capacitor with two dielectrics, as shown in Fig. 3.11.

Tutorials

Tutorials generally follow the last lecture and cover the contents of the lecture. Additionally, they also aim to reinforce materials covered in past topics as well. Tutorial slides are mostly in Japanese with the aim of preparing students toward graduate school/ professional entrance exams. In Japan, these exams are typically given in Japanese, and the questions therein tend to have a particular structure. The tutorials have been designed with this in mind, as a way of familiarizing students with the terminology and approaches used in these cases. The tutorial itself is given in English, and for those students who need it, a translation of the tutorial questions in English is available on the course Moodle.

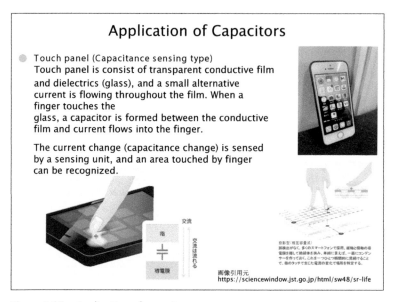

Figure 3.10   Application of capacitors.

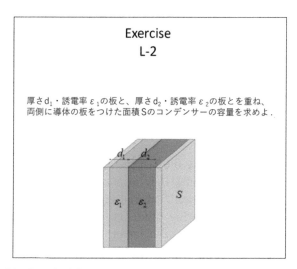

Figure 3.11   Exercise L-2.

The first few slides of the tutorial review the contents of the lecture. Thereafter comes some questions relevant to the topic. Here we show a typical question E-8 (Figs. 3.12–3.15).

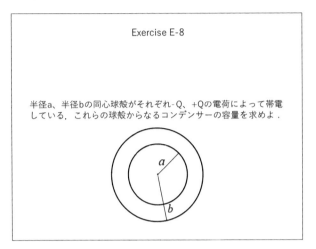

Figure 3.12 Exercise E-8.

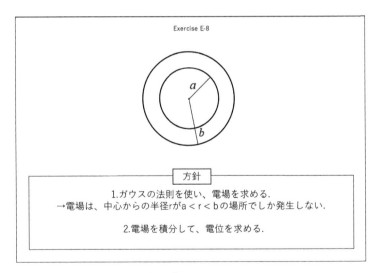

Figure 3.13 Exercise E-8 continued.

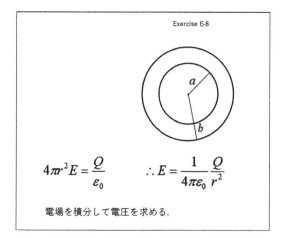

Figure 3.14 Exercise E-8 continued.

Exercise E-8

$$V = -\int_a^b \frac{1}{4\pi\varepsilon_0} \frac{Q}{r^2} dr = -\frac{Q}{4\pi\varepsilon_0} \int_a^b \frac{1}{r^2} dr = -\frac{Q}{4\pi\varepsilon_0} \left[-\frac{1}{r} \right]_a^b$$

$$= -\frac{Q}{4\pi\varepsilon_0} \left\{ \left(-\frac{1}{b} \right) - \left(-\frac{1}{a} \right) \right\}$$

$$= -\frac{Q}{4\pi\varepsilon_0} \left(\frac{1}{a} - \frac{1}{b} \right) = \frac{Q}{4\pi\varepsilon_0} \left(\frac{1}{b} - \frac{1}{a} \right)$$

$$C = \frac{Q}{V} \text{ より、} \qquad C = \frac{4\pi\varepsilon_0}{\left(\dfrac{1}{b} - \dfrac{1}{a} \right)} [F]$$

Figure 3.15 Exercise E-8 continued.

This question not only covers an understanding of capacitance but also requires the students to review Gauss' law and the potential gradient equations. During the tutorial, only the slide with the question is shown, and students are given a certain amount of time to try and solve the question on a piece of paper. During that time, the lecturer and the teaching assistant move around the class checking progress, providing hints, and responding to students' questions. After the given time, the solution can be provided. The students may volunteer to present a solution on the black board and explain their process for the benefit of others. If there are no volunteers, the tutor will demonstrate a solution, explaining each step and responding to questions if any.

Figure 3.16 shows a solution to E-8 from one student.

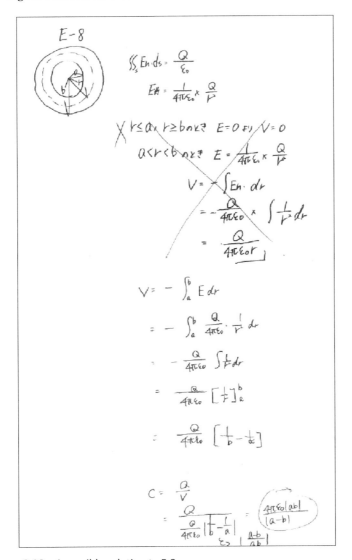

Figure 3.16 A possible solution to E-8.

Midway in working out the solution, they encountered some challenges, which they could overcome with a hint from the lecturer. Afterward they proceeded to the correct solution.

After the tutorial, the solutions to the questions are then provided via Moodle. The slides with the solutions to E-8 are shown in Figs. 3.13–3.15.

In this way, the students can review the content later if there is any part they may have missed, or as they prepare for exams.

The students are also encouraged to utilize office hours provided by the lecturer and the teaching assistant. Typically, these are available immediately after the class and also by appointment via email. In the office hour sessions, students can discuss lecture content, tutorial questions, and any other topics related to physics. Several students have used this opportunity to build paths for their careers through discussions with the professor.

Obtaining credit

As given in the syllabus, the course grade is determined from a midterm exam and a final exam. The exams are, in general, closed books. Students are expected to show an understanding of Maxwell's equations, with an emphasis on Gauss' Law, the Biot–Savart law, and Ampere's law. As such, a majority of the questions are designed to test these concepts and reinforce their understanding.

The exam is usually 70 min long with 12 questions each averaging 10 points. The students can answer as many questions as they can, in any order, within the time limit. A typical exam first page is shown in Fig. 3.17.

The questions are given in English and Japanese. Where relevant, a figure or some hints are provided. The students can answer as many questions as they can manage within the exam time, in any order. They can provide their solutions in either English or Japanese. A typical exam script is shown in Fig. 3.18.

As provided in the question, the student was expected to provide a solution considering both the integral and differential forms of Gauss' law. Here the student has solved in both forms, with labeling and descriptions in Japanese. The lecturer then grades the exam and gives a total mark, which will constitute the final grade.

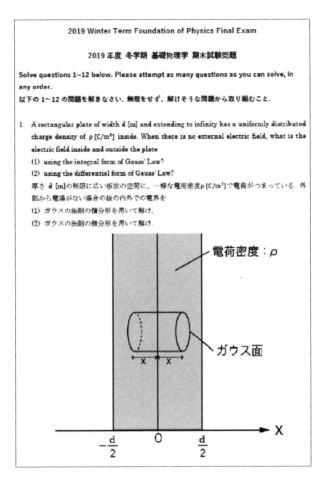

Figure 3.17 Physics final exam.

For students who fail to reach the passing grade, the professor schedules a meeting to gain an understanding of the challenges that the student could not overcome.

Tutor's Impression

I have been tutoring this class for Prof. Okano for the last 5 years, since joining the Master's program in the ICU. Over the years, the distribution of students each year has been different in terms of academic year, background in physics, and future goals. As such each year, Prof. Okano strives to adapt the teaching approach so that each student gets the most benefit from the class regardless of their starting background. As the tutor, my role is to facilitate this transfer to the students and has required similar adaptation. I have also learned by example the essential aspects of teaching a course in physics in liberal arts. My experience watching Prof. Okano's approach is to present physics without its "frills," meaning present physics in "common" language. So far, I can point to a number of approaches I have watched Prof. Okano utilize.

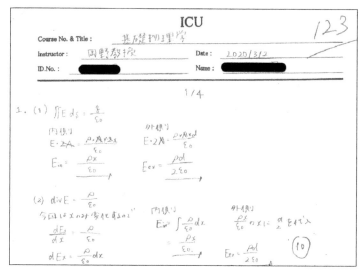

Figure 3.18 Typical exam script.

First is the "low-down" approach. In this approach, the goal is to give the appearance of a small gap between the professor (and tutor) and the student. Rather than instructors appearing as "know-all" superiors, they should present as those with just a bit of extra knowledge, but who are also still in the process of learning more. A typical example, I have seen Prof. Okano use in the very first lecture, is to casually say he still does not understand the real origins of the fundamental permittivity constant ε_0 (I asked and it turns out he has a pretty good idea!). This seems minor during the lecture, but something interesting happens in a later lecture when the permeability constant μ_0 is introduced. Students generally ask if there is an explanation for this constant. I believe if the professor had not presented this "weakness" in the first lecture, students would assume whatever he says is absolute and not question anything else. As the tutor, I have been encouraged to show these "weak points" to encourage students to be able to question solutions and approaches and not consider my presentations as absolute. This tends to be challenging as one usually wants to showcase their masterful knowledge of the subject. Prof. Okano always reminds me of the goal, which is to promote free thinking. Thus, in as much as there may be one final solution to a problem, we need to present our solution in the most flexible way, to emphasize that there are at least a few other approaches that could lead to this solution, and that students should explore and not "cram" the one or two approaches we presented. Then during presenting solutions, I tend to use words and phrases such as "Here is my approach," "Here you could ...".

Another approach is the use of analogies. This seems to help students in imagining the situation in more familiar terms rather than the actual physics. As they gain more knowledge, they transition to an understanding of the concept itself and its associated physics terminology. A typical analogy I have experienced from Prof. Okano in this course is the "Sprinkler and Plastic Bag" analogy representing the electric charge, its flux, and the Gaussian surface. We imagine a garden sprinkler spraying water in a garden in all directions. Our goal is to find the rate of water flow into the sprinkler. One way we could measure this is to cover the sprinkler with a spherical plastic bag and collect all the water sprinkles that hit the bag in a given time. If we measure the total water in the bag, we can figure out the flow rate into the sprinkler head. Similarly, for a collection of charges, to know the total charge, we enclose the charges by a Gaussian surface, which intersects all the emanating field lines. This analogy, though not perfect, is easier to imagine in concrete terms more than the abstract concepts of field lines and Gaussian surfaces. As the tutor, I will repeat the same analogy at the introduction of Tutorial 1, then use the concepts to help remember the integral form of Gauss' law.

One approach I have experienced is the emphasis on real-world examples. Prof. Okano views an appropriate appreciation of physics in the real world as one of the key milestones in the course. As such he will include, in every lecture topic, a reference to how the concept of physics discussed is applied in the real world. For example, the parallel-plate capacitor appears in the first three lectures on Gauss' law, potential, and capacitance, and then its applications are discussed extensively, for example, answering why capacitors tend to have cylindrical shape, capacitors in condenser microphones, and in touchscreens. In a similar manner, magnetism and induction lectures are followed by extensive discussions of magnetic materials in hard disk drives and induction cookers. I have adopted the same strategy to present tutorial solutions by adding some examples from the real world. For example, after students derive the formula for parallel capacitors using Gauss' law, we then discuss how to make capacitance larger. Students contribute that we can increase the area or reduce the separation distance. Here we then discuss some current research on so-called high-k materials and the concepts of super capacitors.

Another interesting real-world example discussed in tutorials is that of thick versus thin wires for carrying current. We solve a question (Figs. 3.19 and 3.20), which shows that for the same volume of wire, stretching it into a thinner dimension increases its resistance. Here, I then give examples of how wire diameter is larger for high-current wires bringing electricity to the homes but is smaller for household appliances.

Exercise E-12

A wire of length L and cross-section area A has resistance R. What will be resistance of the wire if it is stretched to twice its original length?

Assume the density and resistivity of the material do not change when the wire is stretched.

Figure 3.19 Exercise E-12.

Yet another example used in the tutorials is how Ampere's law led to the design of a coaxial cable used, for example, to connect to television antennas (Figs. 3.21 and 3.22). Here, Prof. Okano's goal is to demonstrate the everyday nature of physics and give some concrete or tangible examples that the students can physically interact with in their homes. Ampere's law then ceases to be just about a path integral but something real that the students can see and feel.

Exercise E-12

Twice the length $L_{NEW} = 2L$ Since volume is the same $A_{NEW} = \frac{A}{2}$

$$\begin{aligned} R_{NEW} &= \frac{\rho L_{NEW}}{A_{NEW}} \\ &= \frac{\rho 2L}{\frac{A}{2}} \\ &= \frac{4\rho L}{A} \\ &= 4R \end{aligned}$$

Figure 3.20 Exercise E-12 continued.

Exercise E-22

A solid conductor with radius a is supported by insulating disks on the axis of a conducting tube with inner radius b and outer radius c as shown in the figure. The central conductor and tube carry equal currents I_1 and I_2 in opposite sections of directions. The currents are distributed uniformly over the cross each conductor. Derive an expression for the magnitude of the magnetic field
(a) At points outside the central solid conductor but inside the tube ($a<r<b$)
(b) At points outside the tube ($r>c$)

Figure 3.21 Exercise E-22.

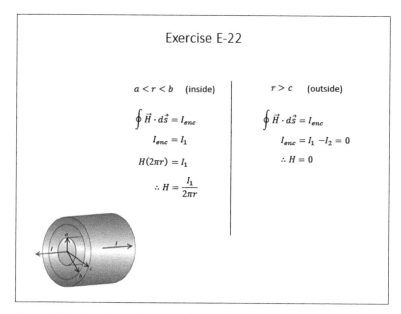

Figure 3.22   Exercise E-22 continued.

A major goal of Prof. Okano in this course, and in general, is what I would describe as demystifying nomenclature. It also turns out to be the biggest role for the teaching assistant, in my opinion. Physics textbooks, while very well written, usually assume some basic background knowledge, for example, in calculus. As such the student is expected to have at least a basic understanding of some symbols used in calculus, for example what an integral is or a delta. Nevertheless, the distribution of students who join the class shows that some may not have ever taken calculus and, therefore, have no background about some of what would be considered basic understanding of the symbols. It is to these students that we intend to provide a foundational understanding of electromagnetism. It is, therefore, our role to also present an understanding of the basic nomenclature. Prof. Okano describes it in a way that by the end of the course, the students are not intimidated by the nomenclature. They also develop an appreciation of it such that in the future, should they encounter it in unfamiliar circumstances, for example as a government policy maker or patent attorney, they should be able to think critically about the concepts without being distracted or confused by the "fancy" appearance.

Gauss's law

Differential form of Gauss's Law

Consider a small cubic. The area of surface at position x is Δy Δz,
So the integral $\int E_n\,dS$ of this surface is

$$E_x(x,y,z)\Delta y\Delta z.$$

And the integral of the surface at position x+ Δx:

$$E_x(x+\Delta x,y,z)\Delta y\Delta z = \left\{E_x(x,y,z)+\frac{\partial E_x}{\partial x}\Delta x+\cdots\right\}\Delta y\Delta z.$$

And hence,

$$E_x(x+\Delta x,y,z)\Delta y\Delta z - E_x(x,y,z)\Delta y\Delta z \cong \frac{\partial E_x}{\partial x}\Delta x\Delta y\Delta z$$

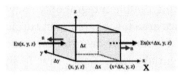

Figure 3.23 Gauss' law.

Gauss's law

Same for y and z axis, so

$$\frac{\partial E_y}{\partial y}\Delta y\Delta z\Delta x, \quad \frac{\partial E_z}{\partial z}\Delta z\Delta x\Delta y$$

Apply Gauss's law to surfaces of the small cubic

$$\iint_S E_n\,dS = \left(\frac{\partial E_x}{\partial x}+\frac{\partial E_y}{\partial y}+\frac{\partial E_z}{\partial z}\right)\Delta x\Delta y\Delta z = (\nabla\cdot\mathbf{E})\Delta x\Delta y\Delta z = (div\mathbf{E})\Delta x\Delta y\Delta z$$

The volume of the cubic is ΔxΔyΔz, so the density of charge in the cubic is

$$\frac{q}{\Delta x\Delta y\Delta z} = \rho(x,y,z)$$

Therefore,

$$div\mathbf{E} = \frac{\rho}{\varepsilon_0} \quad \text{(differential form of Gauss's law)}$$

Figure 3.24 Gauss' law continued.

For this reason, tutorial sessions are devoted to this demystification of the symbols and nomenclature. In the first tutorial on Gauss' law, we have to explain the concept of an integral as a way to calculate the path length, area, or volume by considering the sums of many small subdivisions of regular shape. From there we can then explain why Gauss' law uses the double integral "$\iint$," i.e., calculating the surface area of the Gaussian surface using many small subdivisions. Here we then show that if a convenient regular shape is chosen for the Gaussian surface, then there is no need to do the actual integration, rather just use the formula to calculate the surface area of that regular shape. For example, we solve for the electric field from a point charge and show that a sphere is a convenient shape with a surface area of $4\pi r^2$. In this way, the rather mysterious looking double integral only conveys that one needs to consider a surface area for the Gaussian surface.

In a similar fashion, we use the tutorial to present in simple terms the differential form of Gauss' law. To show that the "up-side-down delta," ∇ or *div*, is no mystery, we undertake the derivation of the equation from the now familiar integral form of the law. Here we also introduce the Taylor series expansion as a way of function approximation. Then the solution is combined to show that it contains partial derivatives in the x, y, and z directions. Since they are partial, instead of d we use the del symbol to indicate that each derivative is only part with respect to one axis. Then the sum of all these parts is what is referred to as the divergence of the electric field E, expressed as div or as the up-side-down delta. Again, the purpose of the exercise is to show that the symbol div is derived from the basic symbols of differentiation. Should the students encounter a div in the future, for example in Einstein's gravity equations, it is no longer such a mysterious concept. In a similar manner, we explain the Laplacian and rotation.

Another impression I got was the use of critical thinking elements in understanding topics and concepts. This was typically applied in working out derivations of some fundamental formulas. The derivation is presented from its goal or purpose, and the question behind it. Students are presented with the assumptions and concepts that underpin the derivation. At the end, the implications and inferences are presented. In general, Prof. Okano then encourages students to have this type of critical thinking or questioning attitude, and to practice it in their problem solving, hoping that the skill gets transferred to a more general application. Because this is explicitly encouraged, as the tutor I have seen students get more comfortable with questioning assumptions during problem-solving exercises. A typical example is why we assume continuity at boundaries, for example, as we solve for the electric field inside and outside a charged sphere.

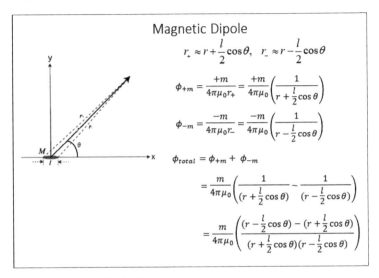

Magnetic Dipole

$$r_{+} \approx r + \frac{l}{2}\cos\theta, \quad r_{-} \approx r - \frac{l}{2}\cos\theta$$

$$\phi_{+m} = \frac{+m}{4\pi\mu_0 r_{+}} = \frac{+m}{4\pi\mu_0}\left(\frac{1}{r + \frac{l}{2}\cos\theta}\right)$$

$$\phi_{-m} = \frac{-m}{4\pi\mu_0 r_{-}} = \frac{-m}{4\pi\mu_0}\left(\frac{1}{r - \frac{l}{2}\cos\theta}\right)$$

$$\phi_{total} = \phi_{+m} + \phi_{-m}$$

$$= \frac{m}{4\pi\mu_0}\left(\frac{1}{(r + \frac{l}{2}\cos\theta)} - \frac{1}{(r - \frac{l}{2}\cos\theta)}\right)$$

$$= \frac{m}{4\pi\mu_0}\left(\frac{(r - \frac{l}{2}\cos\theta) - (r + \frac{l}{2}\cos\theta)}{(r + \frac{l}{2}\cos\theta)(r - \frac{l}{2}\cos\theta)}\right)$$

Figure 3.25   Magnetic dipole.

An example of this critical thinking analysis is the derivation of the magnetic potential (Figs. 3.25 and 3.26). The goal or purpose is to derive a formula for magnetic potential, similar to electric potential. The concepts are the ideas of potential already covered in electricity topics, now with the aim of understanding how they apply to magnetism, considering that magnets are dipolar, compared to monopoles of electric charge. The question then becomes "What is the magnetic potential experienced at a point *P* a distance *r* from a magnetic dipole?" The available information includes the distance to point *P*, which is r; the angle from the dipole to the point *P*, which is theta; the distance between the north and south poles of the magnetic dipoles, *l*; and the magnitude of the magnetic charge at each pole, which is *m*. Several assumptions are made in this derivation. First is that each pole can be treated independently, as if it were a monopole. This allows the use of the same concept of potential from electricity. Another assumption is that the distance between the poles is very small such that theta can be treated as being equal at the north pole as it is at the south pole. The third assumption is that the distance between the poles, *l*, is much smaller than the distance to the point *P*. Combining these concepts and assumptions, the solution can then be derived (see slide). Implications of the derived formula are then discussed regarding the influence of the magnetic field moment *ml* and orientation theta on the magnetic strength at some distance from the magnet.

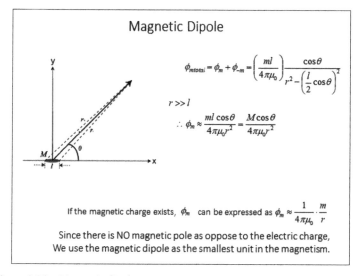

Figure 3.26   Magnetic dipole.

I have observed that as Prof. Okano repeats this approach, students become more confident in analyzing their solutions and questioning the assumptions made. This then tends to present very difficult questions, which in some cases I had not thought about! This has led to some interesting discussions with students, and sometimes afterward, between Prof. Okano and me on some fundamental ideas of physics.

As he promotes these ideas of critical thinking, there is a particular lecture on energy where Prof. Okano then sparks discussions on the future of energy sources, particularly nuclear energy. This is one of my favorite lectures because of the "heated" debate that ensues. As is well known that Japan has the most recent nuclear-energy-related accident after the great earthquake. There tends to be a strong drive in the media to denounce the use of nuclear energy. As Prof. Okano puts it, some reasoning may be correct but some of it is not. His goal, therefore, is to encourage students to actually question the information being presented to them. In the lecture, he sparks the debate by saying something like nuclear energy is essential, if not necessary, and gives some reasoning behind this. Soon enough students challenge the idea. However, in these responses, other students begin to question the reasoning and the assumptions, leading to a very engaging back and forth debate. At the end, Prof. Okano then shows the value of critical thinking in analyzing popular opinions and communications, which students may have simply consumed without paying much attention.

Student's Impression (Psychology Major, No Background in Physics)

Through this course, I became greatly familiar with university-level physics. Before taking this course, I assumed that university-level physics must be really difficult because the calculation process seemed so complicated. This course does not force students to memorize complicated calculus. Rather, the professor gives a detailed explanation of the phenomena, and we need to master only simple calculus. Memorizing complicated calculations is probably important, but not the nature. Especially for those who want to learn physics as part of liberal arts, complicated calculations are not necessary.

(Figure 3.27) The most impressive moment through this course was when I found that the capacitor's formula can be derived from Gauss' law. This is when I realized that physics is not a memorizing-required subject. In my high school time, I was forced to memorize the capacitor's formula without knowing why. However, once I remembered Gauss' law, I became able to apply this to many physical phenomena!

(Figures 3.10, 3.28, 3.29) As shown earlier, examples used in the class were quite familiar topics such as "why getting shocked." I simply thought that we would be shocked if I touch a single line. Those familiar examples made me feel that physics is around my daily life. I also did not know about the mechanism of the touch panel. When I found what the touch panel has to do with electricity, it made sense that my smartphone does not respond with gloves.

Capacitance

⬤ Capacitance: Amount of electric charges stored when 1V is applied

$$\iint_{S} \mathbf{E} \cdot d\mathbf{S} = |\mathbf{E}| \cdot \iint dS = |\mathbf{E}| \cdot A = q/\varepsilon_0$$

$$|\mathbf{E}| = \frac{q}{\varepsilon_0 A} = \frac{\sigma}{\varepsilon_0} \qquad (\sigma : \text{Charge density})$$

In this case, different kinds of electric charges (plus and minus) are stored between two electrodes
-Between two electrodes, the magnitude of electric field is constant at any point
-At rest of the area, the magnitude of electric field is zero
Difference of electric potential (voltage) between electrodes can be expressed as :

$$V = E \cdot d = \frac{qd}{\varepsilon_0 A}$$

From the definition $C = Q / V$, $\qquad C = \varepsilon_0 \dfrac{A}{d}$

(capacitance of a flat plate capacitor)

Figure 3.27 Capacitance.

Why are they sitting so nicely?

⬤ Because they are sitting on **a single line.**
There is no difference in electric potential between two close points in a single line
→No current is flowing

If they touch two points with different potential, they would get shocked

---

Figure 3.28 Birds sitting on single power line without being electrocuted.

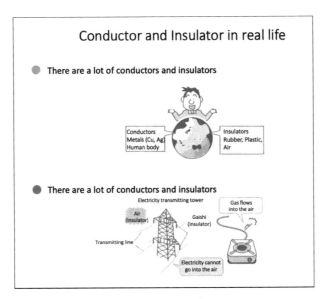

Figure 3.29 Conductor and insulator in real life.

(Figure 3.30) In this lecture, energy and power generation, transmission, we discussed whether we must abolish the nuclear power system or not. Some students think that abolishing this and changing their lifestyles into minimum ones are better. I have never imagined that physics class can be interactive.

In addition, office hours helped me to understand the content of this course. I went to Prof. Okano's office three times a week and asked him many questions. Although it took much more time for me to understand the content than other science-majoring students, Prof. Okano repeatedly explained until I completely understood. In the ICU, the distance between students and professors is close. This enables students to learn countless things from professors.

When I took this course, I also took a course called "Neurobiology," in which I learned the mechanism of neurons and how stimuli are transmitted. Electric phenomena involve transmission of stimuli. When the potential difference occurs between outside and inside the cell, specific ions enter the cell because ions are charged. Since I took the physics course, I could understand this mechanism smoothly and I felt that biology and physics are connected! That is one of the moments I felt liberal arts.

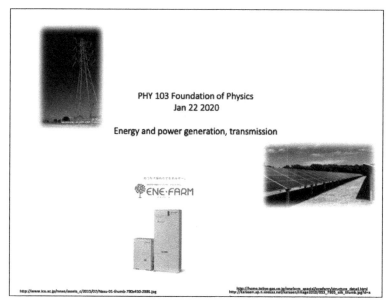

Figure 3.30 PHY 103.

Foundation of Physics Laboratory (PHY105)

Most of the students in Japan were taught to study using textbooks, and it seems they strongly believe that research is an extension of studying based on textbooks. It is true in a sense that some of the subjects are based on textbooks, but it may not be true for all. This course aims to tell the students that studying in the classroom may not be the best way to learn.

One of the topics in this course is "Black Box experiments" in which some of the basic electric parts, such as a resistor, a capacitor, an induction coil, and so on, are connected and mounted in a black box. As you can see, the box is not transparent and so no one can look inside. Of course it has a lid, but it is locked. The only information I give them is "There is either a resistor, a capacitor, and an induction coil, up to two items connected either in parallel or in series, inside the box. Please clarify what is/are inside the box with circuit constant and how they are connected." Most of the students have already learned the electric circuit and know how to calculate the combined resistance/impedance. But even the outstanding students cannot even start the experiments. Why? Because they were never taught how to deal with unknown circuits!

They usually struggle for a few weeks (more than 10 hours), and after I hear their conclusion at the presentation slot, I give a key to open the lid. I always enjoy seeing their reaction.

From my point of view, this experiment is a miniature version of research, because textbooks do not play an important role and the physical knowledge to understand the phenomenon is not at all complicated. In other words, "intransparent box" and "lid" make a psychological barrier so that they cannot answer easily. It can be proved when I ask them to confirm the result after opening the lid; all students can easily confirm what is inside. In addition, the most difficult circuit to be clarified is surprisingly an individual resistor. Can you believe this?

Student's Impression (Psychology Major, No Background in Physics)

"Foundation of Physics Laboratory (PHY105)" is one of my favorite courses in the ICU because of its black box experiment. Through the course, I learned the nature of experiment and was given the opportunity to think about what a genuine researcher is.

An ordinary physics experiment would be like this: students measure voltage, resistance, or current, knowing what type of electrical circuit is. Nevertheless, this course forced me to measure those without knowing what electrical circuit is! I initially did not know why I had to do the experiment in that way. However, as I made progress, I found one thing: Results can often be skewed depending on what the experimenter wants to see. In the experiment world, it is essential to interpret data objectively. A bit of subjective perspective skews the interpretation of data. Thus, how one can be objective toward raw data is really significant.

Another thing I learned through this experiment is this: In the real experiment, we do not know what results are. In other words, measuring an electrical circuit with knowing is merely confirmation. Rather, measuring without knowing what is inside the black box is so close to the real experiment. There are always errors and differences between theoretical and empirical data. I had a quite hard time predicting which type of electrical circuit is inside the black box since the graph based on the raw data did not seem to be applied to any theoretical ones. Though, considering errors, I eventually succeeded in predicting the circuit!

Now, I am studying to be a medical doctor. In the medical world, there are a bunch of phenomena to be revealed. As a doctor, I would like to do research and experiment using what I learned in this course.

Experimental Approach to Natural Science

Looking at this picture, you may think "Are they just playing?" No, this was taken during a physics class. On this day, students in the ICU launched a rocket made from plastic bottles on the lawn.

This class is "Experimental Approach to Natural Science," one of the general education classes available for anyone in any grades regardless of their majors. This class aims to make students find brand new interests in academic fields that they have never been exposed to.

The class is held three times a week, offering 10 classes for chemistry, physics, and biology in order. This picture was taken in the ninth physics class.

Students pour water into the plastic bottle and fill it with air. After they fix the plastic bottle rocket to the shooting table and open the valve, compressed air forces out water, making the rocket fly. Although flying a rocket itself is a play, students investigate how to fly it further, considering how much water is suitable and blending sticky liquid (such as honey or laundry starch) with water. This is the very science.

At the beginning, Prof. Okano drew a graph on a black board and gave an explanation. The x-axis is the amount of water poured in the plastic bottle, and the y-axis is the predicted flying distance. If the plastic bottle is filled with water, there is no room for air and the flying distance would be zero. If there is no water inside the plastic bottle, the distance would also be zero. From those, it can be inferred that there is the best water amount for the farthest flying distance.

Figure 3.31 Launching of a rocket.

After discovering the best amount of water, students investigate what would happen if they blend water with sticky liquid such as laundry starch and honey. They determined the experiment plan.

Let us go outside and do the experiment. One of the students in the class walked toward the landing place of the rocket and measured the distance by steps.

At first, the students poured water up to two-thirds of the plastic bottle. They then blasted off the plastic bottle rocket three times, but the flying distance was about 30 steps. Next, they poured water up to half the bottle. The flying distance was more than 100 steps in fourth and sixth times! There can be the best water amount around here.

In the seventh time, the students filled a quarter bottle of mixed water with the same amount of laundry starch. Surprisingly, the flying distance was 130 steps. "I didn't expect that the rocket would fly, because it was hard to pump air," said a senior student.

In the eighth time, students used honey. The student who works at a bar blended water with honey well and the distance was 122 steps. The students used potato starch next, and the distance was 105 steps. "Do you guys see the story?" said Prof. Okano.

After flying the rocket three more times, the students went back to the classroom and drew the data of water amount and flying distance. They wrote a report based on the collected data, but it is not just that.

In the next experiment class, flying the film case rocket, the students have to fly it more than a graduate student. Otherwise, they will fail.

On that day, the students gathered in the campus hall. The film case rocket flies in the following way: pour water, put a bath bomb (in order to change normal bath water to hot springs!) in the film case, and close it; then carbon dioxide bubbles are generated from the bath bomb, and its pressure makes the rocket fly.

The graduate student did the model experiment, and the distance was about 12 m. Other students followed him.

Like the plastic bottle rocket, the students can adjust the amount of bath bombs and water. However, they found it hard to fly the rocket more than the graduate student. Many ended up unexploded, and 11 out of 18 failed to surpass the graduate student's record. For those who failed, Prof. Okano would hold the make-up class.

Prof. Okano came up with this course since he had been disappointed that a number of students do not have a positive impression toward physics. He reached this style by considering neither dangerous nor so fast materials.

"It is hard for high school students to become interested in physics because they think physics is boring and not related to their daily lives. In order to make them understand that physics is not so difficult, I have to do the class as if playing. Once they know that physics is understandable, they will learn on their own," said Prof. Okano.

One freshman student who had taken a science course in high school, but was not good in physics, said "I thought physics is just formula. I did not know those experiments are also physics! I should have been more motivated to learn physics in high school."

One of the students who passed the film case rocket came to the make-up class. "You see," Professor Okano smiled.

Student's Impression (Psychology Major, No Background in Physics)

I took this course when I was in the second year. I was one of the students who thought physics was so complicated. The reason why I took this course was that I had to take some courses from the natural science department. At that time, of the three science subjects, I disliked physics the most since I thought physics requires a huge capacity of the brain to memorize formulas and equations and I was quite reluctant to participate in the first lecture. However, the lecture was far different from what I expected. "Flying the rocket further? Sounds interesting!" I thought.

Through the course, I was not forced to memorize any equation. Instead, I was forced to find regularity based on collected data, which is more empirical. Usual physics classes are conducted based on equations: the deductive method. However, this course adopts the inductive method, and I felt interested in discovering regularity. And when the collected data and my prediction matched the existing equation, I felt very excited. After that, I could not stop thinking about various phenomena. This course not only eliminated my assumption that I am bad in physics, but also taught a scientific way of thinking.

Concluding Remarks: Education in Japan for the Next 50 Years

The educational system in Japan should be changed drastically as the present system does not yield internationally competitive students from Japan. At the same time, as we can imagine, education, in general, takes a long time until it can be reflected to our society. Our aim is to start changing the Japanese educational system now. If not, we will be far behind the globalization in next decades, compared to the countries having sophisticated educational systems.

We will be glad if this book can be a guide to Japan's future educational system.

Index